TREE FALLER'S MANUAL

TECHNIQUES FOR STANDARD AND COMPLEX TREE-FELLING OPERATIONS

CSIRO PUBLISHING

FOREST, WOOD, PAPER & TIMBER PRODUCTS INDUSTRY

© ForestWorks 2011
Reprinted 2011
Reprinted with corrections 2013
Reprinted 2014

National Library of Australia Cataloguing-in-Publication entry

   Tree faller's manual : techniques for standard and complex tree felling operations/by ForestWorks.

   9780643101548 (pbk.)
   9780643101746 (epdf)
   9780643102286 (epub)

   Tree felling.
   Tree felling – Standards.
   Trees, Care of – Standards.
   ForestWorks.

   363.119634

Published by
CSIRO PUBLISHING
150 Oxford Street (PO Box 1139)
Collingwood VIC 3066
Australia

Telephone: +61 3 9662 7666
Local call: 1300 788 000 (Australia only)
Fax: +61 3 9662 7555
Email: publishing.sales@csiro.au
Web site: www.publish.csiro.au

Illustrations by David McElvenny and Kath Ware, Workspace Training, NSW
Front cover photo courtesy David McElvenny
Photo on page 15 courtesy ForestWorks
Photo on page 53 courtesy Greg Howard

Set in 11/14 Adobe Gill Sans
Cover and text design by James Kelly
Typeset by Desktop Concepts Pty Ltd, Melbourne
Printed in Australia by Ligare

CSIRO PUBLISHING publishes and distributes scientific, technical and health science books, magazines and journals from Australia to a worldwide audience and conducts these activities autonomously from the research activities of the Commonwealth Scientific and Industrial Research Organisation (CSIRO).

The views expressed in this publication are those of the author(s) and do not necessarily represent those of, and should not be attributed to, the publisher or CSIRO.

Acknowledgements

This book is based on the *Chainsaw Operator's Manual: The Safe Use of Chainsaws* 6th Edition 2005

Technical advisors for this edition:
Andy Cusack
Logging Investigation & Training Association, SA

Bill Towie
Forest Products Commission, WA

Greg Howard
Timber Training Tasmania

Ross Connolly
TAFE NSW – Training and Education Support, Industry Skills Unit – Orange

Trevor Wait
East Gippsland TAFE – Timber Training Co-ordinator – Forestech

Assistance from:
Peter Chaffin – Training for Trees, Qld
Goetz Graf – Tree Management Aust Ltd, Qld
Karl Liffman – Timber Training Creswick, Vic
Barry McGregor – NSW Forest Products Association
Ian McLeod – McLeod Training, Qld
Steve Smith – Steve Smith Chainsaw Training, Qld
Ray Stone – Chainsaw Accreditation Safety Training, NSW

Contents

1. Introduction   1
   Compliance with licensing   1
   Tree-felling categories   1

2. National competency standards   3

3. Professional attitude   4

4. Workplace safety   5
   Risk assessment for tree fallers   5
   Equipment, competence of faller and environment   5
       Personal protective equipment   5
       Your equipment   7
       Other equipment   7
       Competence of faller   9
       The current and predicted weather conditions   10
       The tree and its environment   10
       Safe working distance   10
       Escape route   11
       Felling technique   11
       Before starting the chainsaw – final check!   11
   Safety warning signs used for felling operations   12
   Hazardous trees   13
       Typical examples of tree hazards   13
       Typical examples of other hazards   13
       Risk control process for trees with identified hazards   14

Tree felling   15

5. Theory of felling   16
   Importance of directional felling   16
   Considerations before felling each standing tree   16
       Natural lean of tree   17
       Weight distribution of crown   18
       Which tree to fell first   18
       Check for defects   18
       Intergrowth with adjoining trees   18
       Contact with adjoining trees   19
       Wind conditions   19
       Hangers or widow makers   20
       Hung-up trees   20
       Open spaces   20
       Special conditions   21
       Terrain   21
       Hazards in the work area   21

6. Preparation at each tree prior to felling   22
   Clean around base of tree   22
   Prepare escape route   23
       Escape route   24

7.   Standard tree-felling techniques   25
      The scarf   25
            Function of the scarf   25
            Types of scarfs   25
            Features of a good scarf   26
            Depth of scarf   27
      Standard scarf   27
      Other types of scarf   28
            Humbolt scarf   28
            90° scarf   28
            V scarf   29
      Scarf cutting technique   30
            Putting your body in position   30
            Aiming along the desired direction of fall   30
            Achieving scarf cut angles   31
            Size of opening   32
            Wing cuts   33
            Scarf cuts not matched (unequal depth)   34
            Back cut   36
      Back cutting techniques   37
            Method A   37
            Method B   38
      Hingewood   39
            Hingewood for the side lean   39
      Splitting trees   40
            Tree species   40
            Causes of splitting   40
            Felling procedures to minimise splitting   41
      Lifting trees   41
      Wedges   41
            Types of wedges   42
            Using a wedge   43
            Felling bars   43
            As the tree falls   43
            Post-felling checks   44
      Tree with side lean   44
            Tapered hingewood   45
            Tapered hingewood plus wedge   46
      Tree with forward lean   47
            Back release cut – strap technique   47
      Double leaders   48
      Trees with backward lean   49
      Techniques for small trees with the use of wedge or felling bar   50
            Technique 1   50
            Technique 2   52

Complex felling techniques 53

8. Felling hazardous trees   55
    General guidelines   55
    Large trees   55
        Centre scarf   56
    Defective trees   57
        Hollow trees   57
        Stags   58
        Burnt-out trees   59
    Large multiple leader trees   60
    Windblown trees   61
        Felling technique for windblown trees still attached to stump   61
        Tree snapped off above the ground   61
        Root ball is lifted   62
        Bent or stressed trees   62
    Heavy forward-leaning trees with diameter greater than bar length   63
        Strap technique   63

9. Other techniques   64
    Machine-assisted manual felling   64
    Tree jacking   64

Glossary   65
Relevant state authorities and technical standards   66
Technical standards   66

1. Introduction

Tree felling is a high risk activity. Many fatalities and serious injuries have occurred as a result of being struck by falling trees, dislodged tree limbs, or other trees or stags in the area. Most of these accidents are caused by using unsafe felling techniques and/or not following safe work procedures.

This manual is designed primarily for use by timber workers in forest coupes and plantations, but is also a valuable reference for people in other industries who need to fell trees manually using a chainsaw.

Previous editions of the *Chainsaw Operator's Manual* covered both cross-cutting and tree felling. This work is now divided into two separate books, reflecting the division of the content material into chainsaw operations and tree-felling operations:

- *Chainsaw Operator's Manual: Chainsaw Safety, Maintenance and Cross-cutting Techniques*, covering chainsaw maintenance, safe handling, operating techniques and trimming and cross-cutting of logs; and
- *Tree Faller's Manual: Techniques for Standard and Complex Tree-Felling Operations*, covering risk assessment, theory and techniques for manual tree felling.

Compliance with licensing

Compliance with licensing, regulatory or certification requirements may be required in some states and jurisdictions. Please contact the relevant state authority listed at the back of this manual for current requirements.

Tree-felling categories

For the purposes of this manual, tree felling will be described in accordance with the forest and forest products industry national competency standards. The standards prescribe three categories of tree felling:

1. Basic tree felling – trees to be felled are not complex:
    - tree diameter is not more than chainsaw bar length
    - small dimensions relative to local forest size distribution
    - single stem or non-complex multi-stem
    - lean and weight distribution does not create a complex situation to assess or fell
    - no excessive lean
    - no visible damage and/or defect
    - species and growth conditions not prone to splitting or twisting during felling
    - terrain and slope does not add significant complexity.

2. Intermediate tree felling – trees with some complexities, which may include:
   - trees of varying dimensions in relation to local forest size distribution
   - single or multi-stems
   - lean and weight distribution which can be adapted to felling direction with the use of wedges and/or control with hingewood
   - diameter of tree greater than bar length
   - limited visible damage and/or defect
   - terrain and slope that may add complexity.
3. Advanced tree felling – complex trees, of any size and condition, which are determined as safe by the faller to be felled and may include the following conditions:
   - trees with larger dimensions relative to local forest size distribution
   - trees with substantial lean
   - trees with damage or defects that require complex felling techniques, which may include multi-legged, hollow butts, culls and stags
   - species prone to free splitting and adverse reactions during felling
   - trees with diameter greater than bar length
   - terrain and slope that adds significant complexity.

2. National competency standards

The information in this manual can be used to support training aligned to the units of competency from the Forest and Forest Products Industry Training Package.

National competencies specify the skill and knowledge requirements for performing particular tasks or job functions in the workplace to the standard expected in the industry.

The units of competency listed below are partly covered by this manual. **Please note that completion of this manual does not constitute competence in these units.** Chainsaw operators seeking accreditation in any of the units below should consult a Registered Training Organisation (RTO). A list of RTOs can be found on the ForestWorks website (see below).

Chainsaw Operator's Manual: Chainsaw Safety, Maintenance and Cross-cutting Techniques

FPICOT2237A	Maintain chainsaws
FPICOT2238A	Cut material with a hand-held chainsaw
FPICOT2239A	Trim and cut felled trees
FPIHAR2207A	Trim and cut harvested trees

Tree Faller's Manual: Techniques for Standard and Complex Tree-Felling Operations

FPICOT2236	Fall trees manually (basic)
FPIFGM3212	Fall trees manually (intermediate)
FPIFGM3213	Fall trees manually (advanced)
FPIHAR3220	Harvest trees manually (intermediate)
FPIHAR3221	Harvest trees manually (advanced)

A copy of the units of competency can be downloaded from www.ntis.gov.au or www.training.gov.au.

For information and advice on learning and skills development please contact ForestWorks, the national skills advisory body for the forest, wood, paper and timber products industries: www.forestworks.com.au.

3. Professional attitude

It is essential that a faller maintains a professional and responsible attitude towards all aspects of tree felling and a duty of care for the environment. Attributes that help to make a safe and competent operator include:

- putting safety first
- careful forward planning and risk assessment
- good problem-solving skills
- maintaining a steady work pace without rushing
- concentrating at all times
- using sound, low-risk techniques
- understanding and taking care of equipment
- caring about the environment.

A competent faller is one who can:

- confidently assess a tree they intend to fell with regard to a range of factors, including the tree's natural lean, presence of dangerous limbs, defects in the tree and the influence of any wind upon the felling operation
- apply the correct felling techniques
- fell a tree in the 'desired direction of fall'
- take steps to maximise resource utilisation
- recognise when it is better to NOT fell a tree.

Figure 1: Think before you act.

4. Workplace safety

Risk assessment for tree fallers

Before starting your chainsaw to fell a tree, it is vital that you think about what you are about to do. Risk assessment is a process of carefully thinking through the steps of the task you are about to do, to ensure that every aspect that has the potential to cause injury or death is covered.

You must consider the following:

- Tree felling can be a demanding task both physically and mentally. The tree faller needs to consider their physical capability to handle the exertion of operating the saw, the terrain of the worksite and their capacity to use axe or hammer and wedges.
- In addition to any physical capability, impairment due to drugs or alcohol, fatigue and general health needs to be factored into any risk assessment.
- Are your felling skills, as the operator, commensurate with the task? Have you been adequately trained and hold the accreditation that may be required in the jurisdictions in which you are working?
- Finally, you must be capable mentally to handle the stress of tree felling and the inherent dangers involved.

Equipment, competence of faller and environment

Personal protective equipment

The Australian Standard *AS 2727 Chainsaws – Guide for Safe Working Practices* lists the items of personal protective equipment (PPE) that a chainsaw operator must use. Remember that PPE is at the lower end of the scale of risk control, and will not guarantee your safety. Proper risk assessment and safe work practices must always be followed.

The items on the next page are recommended when working in a forest environment. Refer to the list of technical standards at the back of this manual for more detailed specifications.

Safety helmet (AS/NZS 1801)

Must be replaced if cracked, damaged or past expiry date. Avoid damage caused by attaching stickers, storing in direct sunlight and contact with solvents. A legionnaire-style flap can be attached to protect the back of the neck.

Eye protection (AS/NZS 1336 and 1337)

Preferably non-scratch and non-fogging. Can be either clear or mesh visors or goggles. Safety glasses should also provide an acceptable level of protection.

Steel-capped safety boots (AS/NZS 2210)

Boots should also have non-slip tread and lace-up for better ankle support.

Cut-resistant trousers or chaps (AS/NZS 4453).

The cut-resistant layers cannot be repaired and the garment should be replaced if it has been cut. The effectiveness of the cut-resistant layers may be reduced over time by the absorption of oil when used regularly.

High-visibility vest/shirt (AS/NZS 4602)

Long sleeves are preferable for sun protection. Some vests are also reinforced with cut-resistant fabric for added protection.

Hearing protection (AS/NZS 1270)

May be either ear plugs or ear muffs. Be aware that ear plugs and some cheap ear muffs may not provide sufficient protection when using larger saws. Check with your dealer for correct level of protection required for your chainsaw.

Safety gloves (optional)

Should be snug fitting and of a hard-wearing, protective fabric. Some gloves have gel cushioning to protect against vibration. Protective wristbands and forearm guards are also available.

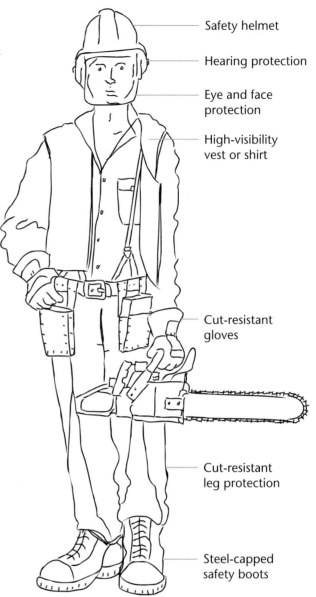

Safety helmet

Hearing protection

Eye and face protection

High-visibility vest or shirt

Cut-resistant gloves

Cut-resistant leg protection

Steel-capped safety boots

Figure 2: Personal protective clothing and equipment.

Your equipment

Some tools and items are also considered necessary for tree felling. They are:

- chainsaw that meets AS 2726
- power head well maintained
- chain sharpened and correctly tensioned
- fuel and bar oil (must have sufficient fuel to complete the felling operation – it may be unsafe to stop the cutting sequence to refuel, leaving a partially cut tree)
- maintenance kit
- axe/hammer and wedges (at least two) must be immediately to hand
- first aid kit (within use-by date)
- communications system or equipment
- adequate supply of drinking water.

Other equipment

Sharp axe

This can be used for many tasks, such as removing bark from a log. The axe should be in good condition, with a forged and tempered head free of cracks and a sharp cutting edge. When not in use, a protective stitched leather cover should be fitted.

The best handles are made of American hickory or Australian hardwood timber. These should be tightly fitted, secured with cross-wedges and/or pinned. Ideally a handle should be straight grained, generally smooth and free of cracks or knots, and have a roughened grip. Linseed oil may be applied to keep the handle in good condition.

Wedges and hammer

At least two wedges are required. These should be made of robust aluminium alloy or lightweight plastic that will not adversely damage a running saw chain if struck. Wedges must be in good condition, free of cracks and mushroomed heads. While not recommended for general use, steel wedges may be useful for fire salvage work.

A suitable hammer of sufficient weight is required for driving wedges. The handle must be of good length, tightly fitted and free of knots, cracks and splinters.

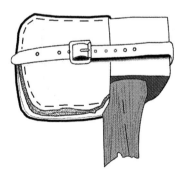

Figure 3: Axe and cover.

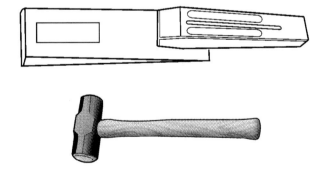

Figure 4: Wedges and hammer.

Cant hook or breaking bar

A cant hook or breaking bar can be of assistance in moving log material and bringing down hung-up trees.

Fuel and oil container

Fuel and oil containers for storage and dispensing of two-stroke mix and chain oil must be of an approved fuel-proof design. Some have 'quick-fill' automatic cut-off fuel pourers.

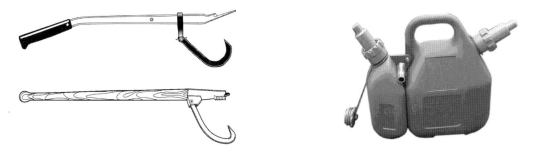

Figure 5: Cant hook and breaking bar, and fuel and oil container.

First aid kit

A fully equipped first aid kit must always be provided at the workplace, and a personal kit must be readily accessible by the operator. Among the contents will be large dressings for lacerations. Sun protection cream must also be available as necessary. The Australian Standard AS 2727 recommends the minimum content of the kit.

A faller must carry on his person, a prepared wound dressing No. 15 while felling.

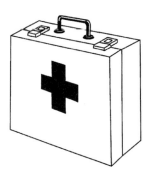

Figure 6: First aid kit.

Fire control equipment

Fire control equipment including a rake-hoe, fully charged hand pump knapsack-type sprayer and fire extinguisher are required during the fire season.

Figure 7: Fire control equipment.

Tool belt

A tool belt is worn for easy carrying of logger's tape, wedges and such small tools as a chainsaw wrench, file and possibly a log vice. A small first aid kit can also be attached.

Figure 8: Tool belt with logger's tape attached.

Logger's tape

Logger's tape is a specially designed retractable, spring-loaded steel tape, necessary for measuring saw logs when cross-cutting to length.

Chainsaw maintenance kit

The most basic chainsaw maintenance kit is that carried on the operator's tool belt. A comprehensive kit contains the following small tools:

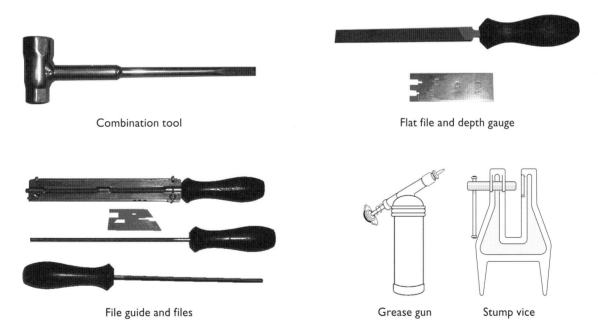

Combination tool   Flat file and depth gauge

File guide and files   Grease gun   Stump vice

Figure 9: Some components of a basic maintenance kit.

Additional items include: wrenches, a flat head screwdriver, wooden scraper, brushes, air filter treatments and cleaning rags. It is essential that all tools and equipment are well maintained in sound condition.

Competence of faller

Assess your competency

• Is it within your capacity to fell the tree safely and competently?

The current and predicted weather conditions

Do not attempt to fell a tree before conducting a situation analysis, and checking for inclement weather, especially high winds. Extreme temperatures, heavy rain and lightning can also increase the risk of tree felling.

The tree and its environment

Assess the tree to be felled and its immediate surrounds

Check the following factors in relation to the trees to be felled and its immediate surrounds.

- Condition of tree.
- Is it a hazardous tree? (Refer to section on complex trees).
- Presence of broken limbs hanging in trees (widow makers) in close proximity. Do not work within the drop zone of a hanging broken limb.
- Hazardous trees in the immediate area including stags, burnt-out or rotted-out trees.
- Intergrowth with adjoining trees.
- Hung-up tree/s in contact with tree to be felled.
- Climbing vines in contact with tree to be felled.
- Presence of other smaller trees or undergrowth that may be a hazard during the cutting process.
- Open space into which the tree can be felled. Do not fell a tree into another standing tree (remove the other tree first).
- Other hazards and activity in the area (e.g. powerlines, traffic).
- Sufficient visibility required for felling (consider distance and light) – no felling at night.
- Desired direction of fall of the tree.
- Safe area into which to fell the tree.
- Tree does not interfere with prohibited area (boundary, buffer zone, etc.).

Safe working distance

- Keep at least twice the length of the tallest tree in the immediate work area between faller and all other personnel and activities.
- Use appropriate warning signs when felling is being conducted.
- Do not enter felling area until adjacent trees have settled.

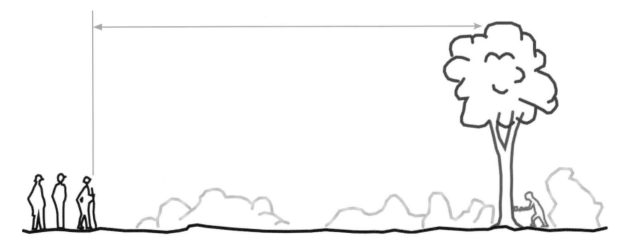

Figure 10: Safe working distance – at a distance which is twice the height of the tallest tree in the immediate work area.

Escape route

- Appropriate route (45° back and away from desired direction of fall) to be free of obstacles for at least 6 metres.
- Always prepare a second escape route if the tree is unsound or deemed in any way hazardous.
- Prepare and use the most suitable and safest escape route.

Felling technique

- Select the felling technique appropriate to the tree being felled and its desired direction of fall.

Before starting the chainsaw – final check!

- There is no person or machinery within two tree lengths.
- Desired direction of fall has been identified.
- Felling technique has been identified.
- The side of the tree you will be at when putting in the release cut has been identified.
- Escape route has been identified.
- Sufficient fuel is available.
- Axe/hammer and wedges are immediately to hand.

> **Remember – if you can't fell the tree safely
> DO NOT attempt it!**

Figure 11: Look up and live.

Safety warning signs used for felling operations

Even small trees have the capacity to kill, or severely injure people, or inflict significant damage to vehicles or machinery. Therefore signs warning that tree felling is taking place should be displayed in all areas where this activity is occurring.

The type, shape, size, wording and colouring of signs displayed may vary significantly throughout the country and depend on state regulatory requirements, state-based codes of practice and enterprise requirements, but must comply with *AS 1319 – 1994 Safety Signs for the Occupational Environment.*

As a general rule signs should be placed on entry roads to the felling area and at a distance from the boundary of the felling area. Signs should provide adequate warning to persons approaching the felling area (in the vicinity of 200 metres from the boundary).

Signs should be kept in good condition and be readable to approaching persons.

> ## Ensure all signs and sign placement meet landowner, state regulatory and enterprise requirements.

The following are examples of some signage that may be used.

Figure 12: Sample signage.

Hazardous trees

Ultimately it is the faller's decision whether or not a tree is too hazardous to fell. Hazards may come from the tree itself, the tree's immediate surroundings or the weather conditions prevailing at the time of felling.

It is vital for the tree faller to assess the condition of a tree and its environment **before** any attempt is made to fell it. Binoculars can assist when making judgements about hazards at height.

Typical examples of tree hazards
- Heavy decay or fire scarring.
- Dead or partially dead.
- Very brittle or has a hollow trunk.
- Broken limbs hanging in its crown (widow makers).
- Natural lean is away from the desired direction of fall.
- Excessive lean.
- Hung-up or has a hung-up tree resting in it.
- Interlocked with adjoining trees.
- Storm or snow damaged.
- Exposed or unstable root system.
- Defective tree in close proximity to tree to be felled (less than two times the tree length).

Typical examples of other hazards
- Thick undergrowth at tree base that cannot be cleared.
- Location that restricts faller's safe movements (boulders, steep banks).
- Trees damaged by previous felling or partially cut trees.
- Inadequacy of and unsuitable condition for wood fibre to ensure safe directional control of falling tree.
- Tree felling when:
    - alone
    - in stormy or windy weather
    - on a fireline, particularly if the tree is alight
    - operator lacks competence for the type of tree or hazard.

Risk control process for trees with identified hazards

Identify hazards (refer to the previous page)

- Make an assessment of the risks from the hazards identified (risk assessment).
- Determine (maybe consult with another faller) whether manual felling is a suitable option or the tree is unsafe to fell.

If manual felling is identified as a suitable option:

- the faller must be competent in felling techniques and have relevant felling accreditation and experience
- a safe separation distance of at least two tree lengths from other workers must be maintained
- adjacent or associated hazards (stags, etc.) are removed first
- the area around the base of the tree must be cleared and suitable escape routes are in place
- communication should be maintained between all operators
- techniques that limit exposure to any hazard are adopted.

If the tree is assessed as being too high risk to fell manually:

- prohibit work near the unsafe tree
- clearly identify the tree to workplace specifications (e.g. mark the tree, tape off the area, record location on coupe or site plan)
- arrange for mechanical equipment or explosives to remove unsafe tree.

Tree felling

5. Theory of felling

A 'competent' faller is one who controls the situation. A 'competent' faller never lets the situation control them and never works outside the limits of their abilities.

Importance of directional felling

A competent faller fells the tree in the desired direction, without injury to themselves or other people, or damage to:

- equipment
- tree being felled
- any sound timber on the ground
- retained trees.

Therefore directional felling is important for:

- own safety
- safety of others
    - don't leave 'hangers' or 'widow makers'
- work pattern
    - ease of extraction by machinery
    - **not** covering up other felled trees
- recovery (utilisation)
    - **not** shattering logs
    - **not** smashing regeneration or retained trees (e.g. crop trees, habitat trees, seed trees).

Considerations before felling each standing tree

When assessing a tree to determine the desired direction of fall, consider:

- natural lean of the tree
- weight distribution of the crown
- which tree to fell first
- defects or damage caused by rot, termites, lightning strike, burnouts, insect nests
- intergrowth with adjoining trees
- wind conditions (strength, direction and type – constant or gusting)
- hangers and widow makers
- hang-ups
- climbing vines

- open space
- special conditions
- terrain (including ground slope, holes, rocks, fallen timber and other obstacles which may obstruct the faller's movement)
- other hazards in the work area.

Binoculars are useful when assessing trees, particularly hazards related to heights.

Figure 13: Assessing a standing tree.

Natural lean of tree

It is difficult to fell a tree against all but a moderate lean. Wherever possible, fell with the natural lean of the tree and the weight distribution of the crown.

An axe/hammer, used as a 'plumb bob', can be used to judge the tree's natural lean.

Weight distribution of crown

Determine which side of the crown has the most weight. Branching or heavy growth on one side may override natural lean and therefore change the direction of fall.

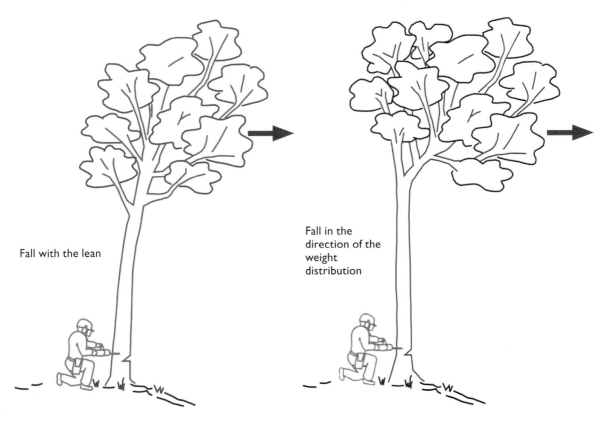

Fall with the lean

Fall in the direction of the weight distribution

Figure 14: Lean of tree and weight of crown.

Which tree to fell first

The faller needs to consider the sequence of felling trees to guard against:

- creating hangers or widow makers
- creating hang-ups
- creating extra work through placing heads of trees or limbs at the base of another tree to be felled.

Check for defects

Defects are assessed visually:

- Look for external scars, cracks, deadwood in crown, burnt sections, insect nests or damage, previous cuts, etc.
- If a tree is suspected of having internal defects, verification of defects can be established from the sawdust obtained when bore cuts are made.

Intergrowth with adjoining trees

Trees may be difficult or hazardous to fell when they are intergrown with adjacent trees or vines are interwoven with adjacent trees.

Contact with adjoining trees

Avoid felling trees that brush against or contact other trees otherwise this may result in:

- falling limbs
- widow makers
- possible hangers
- limbs flinging back towards the faller
- felled tree breaking and falling back towards the faller
- knocking over other trees.

Wind conditions

When assessing a tree to fell, possibly the most dynamic and variable factor is the wind condition. Wind has a significant impact on the ability of the faller to effectively and safely fell a tree. To add to the complexity of the process, the wind also needs to be combined with other factors being assessed such as size of tree and canopy, lean, and condition of tree.

There are a range of factors for consideration:

- direction and strength of wind
- is it constant or multi-directional?
- is it with or against the proposed direction of fall?
- is it strong or light?
- is it gusty and unpredictable?

Constant direction of wind is generally easier to allow for than gusty or variable directional winds. There is a high risk associated with felling in strong winds and felling should not be undertaken.

Consider that if there is difficulty created in the felling situation, by the strength or variability of winds, it is often calmer in the early or later part of the day to undertake the task. Always take into account the form of the crown or foliage and the effect of wind, which can be quite significant as the crown may create a 'sail effect'. There may also be situations where the wind may have a positive effect, for example if there is some backward lean identified or suspected, a following wind may assist the process.

As with any felling activity, if the wind conditions are dangerous or variable be prepared to cancel or postpone the activity.

Remember:

- Wind may be strong enough to overcome the tree's natural lean or the weight distribution of an unbalanced crown.
- You may have to wait for a 'lull' in wind under gusting conditions before and/or during felling.
- Wind velocity is always less at ground level than crown level. (Wind acts on the tree crown in a similar manner to wind acting on a sail of a sailing boat.)
- Avoid using chainsaws in excessively windy conditions.

Hangers or widow makers

Take extra care when felling a tree with a hanger/widow maker – make sure that you do not work under hanging limbs. The first stage of a tree's fall may dislodge the hanger – continue watching during escape. Some trees containing widow makers may be deemed too risky to fell and these trees should be marked and left.

> ## Never work underneath lodged or hung-up trees.

Hung-up trees

- Trees that have lodged together after falling should be marked and immediate steps taken for a machine to pull it down.
- Never walk under a hung-up tree, or attempt to fell the tree in which it is hung, or try to knock it down with another tree.
- All hung-up trees must be marked with hazard tape or a suitable sign displaying 'Danger: Hang-Up'.
- Ensure all signs and sign placement meet landowner, state regulatory and enterprise requirements.
- Cordon off the fell zone with tape.

Figure 15: Example of a suitable sign.

Open spaces

- Always fell into open space (even small trees in the felling line can be thrown back towards the stump and the faller).
- Create your own open space by working on the felling face whenever possible.
- Avoid felling into other trees, stumps, rocks or logs if possible.
- Remove trees systematically to create space for a clear line of fall.

Special conditions

This includes the presence of fences, powerlines, filter strips, etc.

Terrain

When assessing terrain fallers should consider what effect slope, holes, rocks, stumps, fallen timber or other obstacles may have:

- on their movement around the tree during the felling process
- on their ability to access and use their escape routes effectively
- on what may happen if the tree hits any obstacle as it falls or lands.

Where any obstacles are present that may have an effect on the felling outcome, fallers may have to change their desired direction of fall to avoid these obstacles or arrange to have obstacles shifted before felling commences. If this is not achievable and the obstacles/effect may pose an unacceptable risk then the tree may be considered too dangerous to fell.

Hazards in the work area

Always evaluate the tree-felling work area for standing hazards that are present or may be created by the felling process. These may include dead or defective trees.

6. Preparation at each tree prior to felling

Clean around base of tree

- Prepare a clean work area of at least 1 metre around the base of the tree.
- Consider removing saplings and small spars along tree-felling line as these may throw or fling material back towards the faller.
- Remove or cut limbs or logs that extend from in front of the tree and back past the stump towards the escape route. This will avoid one end of this material flying up and hitting the faller if the other end is struck by the falling tree.
- Identify escape routes at least 6 metres in length, and prepare the safest route.

Figure 16: Remove trip hazards.

Prepare escape route

The majority of felling accidents occur within 4 metres of the stump. Therefore:

- Identify two escape routes at least 6 metres in length, with a line of retreat 45° diagonally backwards, away from the direction of fall. (There may be cases when two escape routes are not required, e.g. side-leaning trees).
- Clear escape routes of obstacles for at least 6 metres. When nominating the preferred escape route the following aspects could be considered:
  - up slope rather than down slope
  - finish on side away from lean and weight distribution
  - if there is a hanger/widow maker on one side, choose the other side
  - there may be trees on one side in the direction of falling – use the other side.

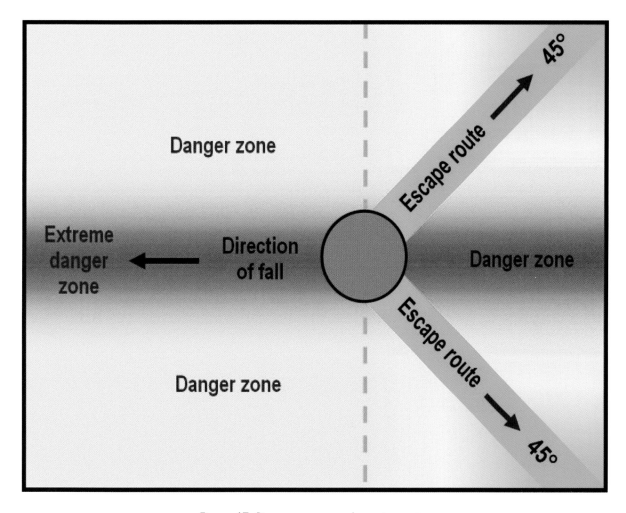

Figure 17: Danger zones and escape routes.

Escape route

The escape route is to protect the faller should any of the following situations occur:

- If the butt kicks up as the tree falls it will generally go straight backwards or to one side.

- If the tree splits up it will slab backwards from the line of fall.

- If the tree snaps in the felling line it will generally come back straight over the stump.

- When felling trees uphill, they may slide straight backwards passed the stump – it is advisable to prepare longer escape routes.

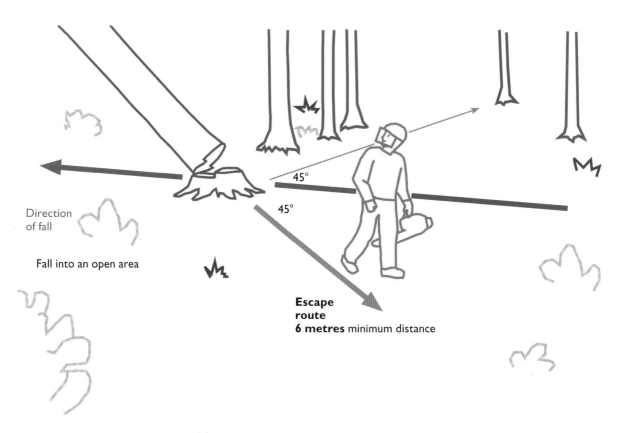

Figure 18: Escape routes.

7. Standard tree-felling techniques

The three components of standard tree-felling techniques are:

- the scarf
- the back cut
- hingewood.

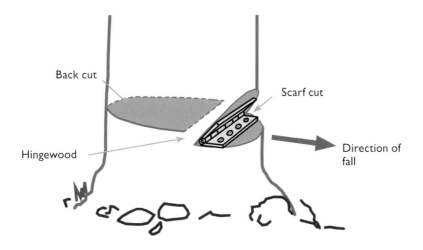

Figure 19: Components of a felling cut.

The scarf

The scarf is the piece of timber that is removed from the front of the tree and faces the desired direction of fall. The opening created by the scarf is primarily to provide a space for the trunk to fall freely into, without hesitation or delay, as the tree tips over in the desired direction. The principal function of the scarf is to guide the tree to its nominated/desired direction. No trees should be felled without the recommended scarf.

Function of the scarf
- Directs tree in desired direction of fall.
- Controls the tree during the arc of fell (allows smooth, steady fall of tree).
- Serves as a means of breaking the hingewood.
- Helps to prevent tree from splitting up.

Types of scarfs
- Standard
- Humbolt
- 90° scarf
- V scarf.

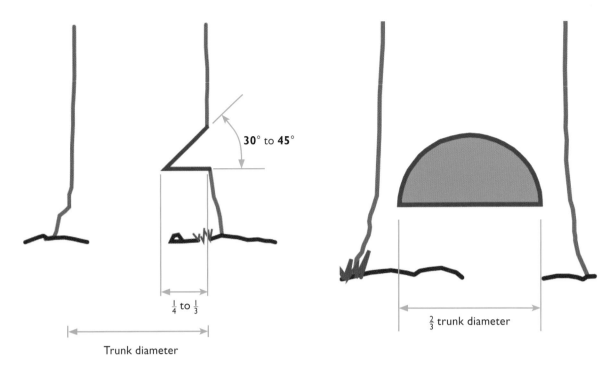

30° to 45°

$\frac{1}{4}$ to $\frac{1}{3}$

Trunk diameter

$\frac{2}{3}$ trunk diameter

Figure 20: Examples of standard scarf.

Features of a good scarf

Direction: must be in direction of desired fell

Depth: generally $\frac{1}{4}$ to $\frac{1}{3}$ diameter of tree

Size of opening: $\frac{2}{3}$ across the front of the tree

Cuts:

- The two cuts should meet without overcutting or undercutting.
- The line where the two scarf cuts meet is called the scarf line (see Figure 21).
- The scarf line must be level. If the scarf line is not level then the hingewood will not be even and the desired direction of fall may be affected.
- The scarf line is at 90° to the intended direction of fall.

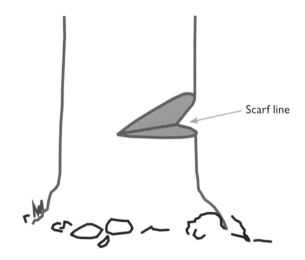

Scarf line

Figure 21: Illustration of scarf line.

Depth of scarf

It is critical that scarf depth does not exceed specifications. This may cause lack of control and significant hazards for effective felling. This may include difficulty in wedging, premature release of the tree, diminished control of hingewood or risk of the tree falling backwards.

Standard scarf

- Most commonly used scarf.
- Consists of horizontal bottom cuts and angled top cuts.
- Both cuts are to finish at the same depth in the tree (no overcuts or undercuts).
- Scarf wood should be removed cleanly.

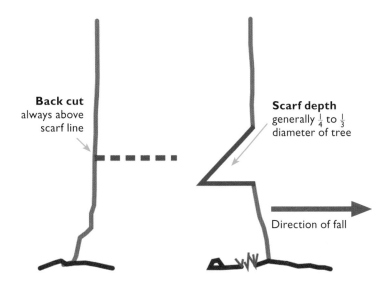

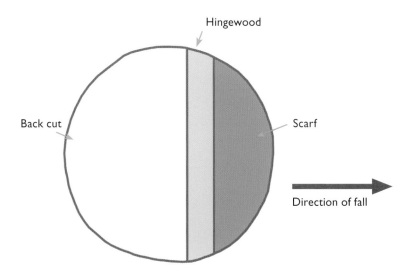

Figure 22: Standard scarf.

Other types of scarf

Humbolt scarf

- Frequently used in saw log operation to maximise recovery.
- Sometimes difficult to cut very low to the ground.
- Can be an advantage when felling uphill. This cut provides extra safety to prevent the tree from sliding back over the stump.

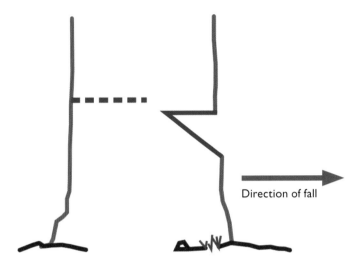

Direction of fall

Figure 23: Humbolt scarf.

90° scarf

- Can be used in trees with very pronounced butt swell and on very small trees.
- Both cuts are relatively easy to match.

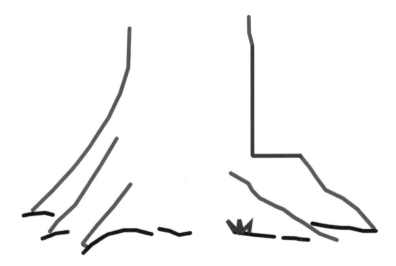

Figure 24: 90° scarf.

V scarf

- Opening as wide as possible.
- This is a fairly difficult scarf as the operator must match the two sloping cuts.
- Advantage is that this scarf gives a very wide mouth opening. Tree falls with control through greater angle.
- Can be used to advantage when felling trees with trunk diameters larger than twice the chainsaw's cutter bar length.

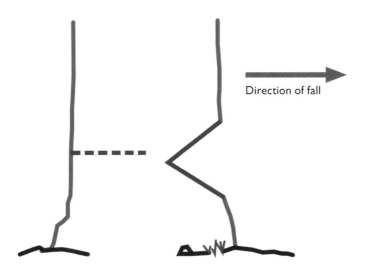

Direction of fall

Figure 25: V scarf.

Scarf cutting technique

1. Putting your body in position
- Support the body and knee against the tree to relieve the strain on the back and to assist in sighting the saw accurately.

Figure 26: Position body.

2. Aiming along the desired direction of fall
- Use the sightline (gunning sight) to align the felling direction.
- The sightline usually extends from the starter cover, up over the top cover and down the sprocket cover.

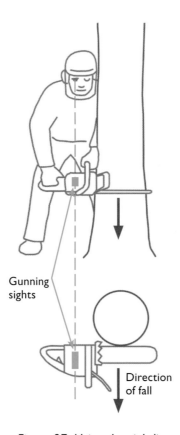

Gunning
sights

Direction
of fall

Figure 27: Using the sightline.

Figure 28: Hand position for top cut angles.

3. Achieving scarf cut angles

- Shift hand position along the front handle, to tilt the saw to the correct angle for sawing the upper cut.
- Handgrip should be placed in the middle of the side handle for correct balance and to keep the saw in a level position for the horizontal cut.

The order of making the cuts in the scarf is not important as long as all other requirements of the scarf are met.

Size of opening

Slope of top cut (or bottom cut) should be 30° to 45°. This is to create an opening to control the tree's fall through as large an angle as possible.

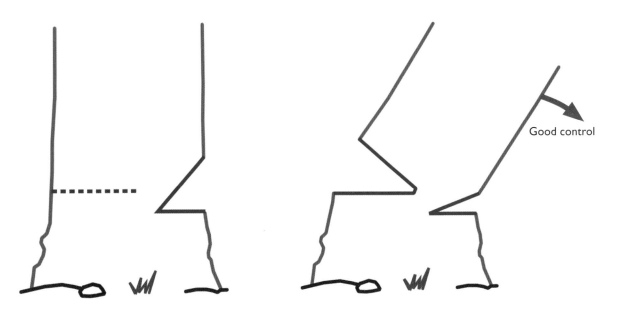

Figure 29: Specified scarf opening 30° to 45°.

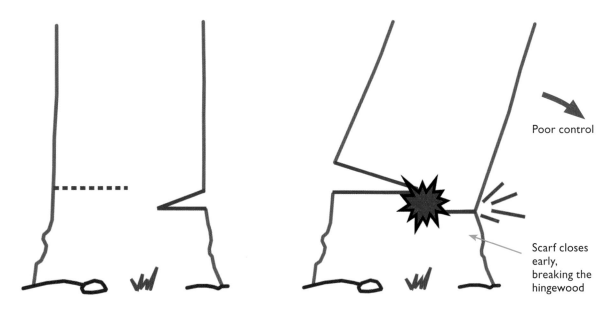

Figure 30: Narrow scarf opening.

If opening is too narrow, then scarf closes soon after the tree begins to fall, thus breaking hingewood too early. A narrow opening can also cause the tree to split/slab or pull and damage wood.

Wing cuts

Wing cuts can be useful to prevent splitting of stem and uncontrolled root pull where the scarf does not completely cover the front of the tree. They consist of a cut either side of the scarf line at approximately 45° into the stump.

Most hingewood breaks when two scarf cuts come together. Immense tension/pressures occur at the stump resulting in the sides of the log being torn or split and/or uncontrolled root pull. This can be hazardous and degrade the butt log.

Wing cuts can prevent this from happening. Wing cuts are completed immediately after completion of the scarf and prior to the back cut. This technique is not recommended for hardwood.

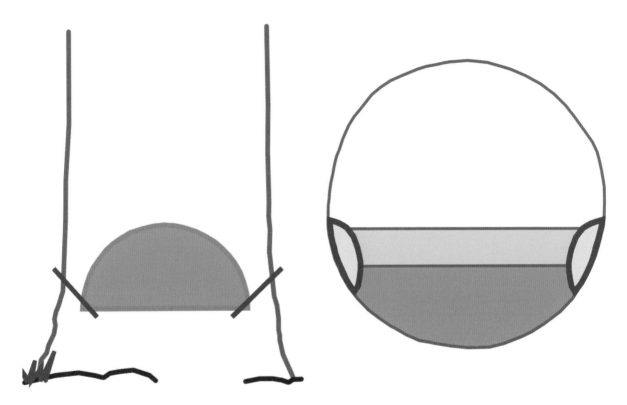

Figure 31: Use of wing cuts.

Technique for wing cuts
- Stand in front of tree.
- Place a cut either side of the scarf line at 45°. The cut is about the width of the bar and down into the stump to about 100–150 mm.

Scarf cuts not matched (unequal depth)

When scarf cuts are not of equal depth (i.e. one cut extends further into the tree than the other) this is often referred to as undercutting (or overcutting), and one or more of the following problems may occur.

1. The tree may sit on the undercut and not fall. To make the tree fall in this instance it will be necessary to continue the back cut, subsequently cutting the hingewood, and unless the tree is being felled in the exact direction of its lean the tree will likely fall in the direction of lean rather than the desired direction of fall.

2. As the tree leans forward and closes the undercut it may break the hingewood prematurely, resulting in loss of directional control early in the arc of fell. Once again, the tree is likely to fall in the direction of lean as opposed to the desired direction of fall. In this instance it is also likely that a substantial amount of wood fibre will be pulled out of the log end, resulting in damage to the end of the log and its quality as a saw log.

3. As the tree leans forward and closes the undercut, the barrel of the tree stops moving. But because the top of the tree has more momentum this places pressure on the barrel of the tree from the extent of the back cut in a vertical line up the tree. If the tree being felled is of a free-splitting species then it is likely the tree will split up from the back cut, with an indeterminate proportion of the tree travelling back past the stump towards the danger zone and the escape routes. This creates an extremely dangerous situation for the faller and has resulted in many injuries and deaths.

4. If the cuts have been matched up on one side of the scarf but not on the other, then in addition to the possibility of one of the above problems occurring it is also likely that when the undercut closes it will break the hingewood on one side of the tree prematurely, resulting in the tree being pulled around towards the side that the hingewood is still holding.

Where scarf cuts have not been matched up, with one cut extending further into the tree than the other, it will be necessary to recut the scarf before commencing the back cut.

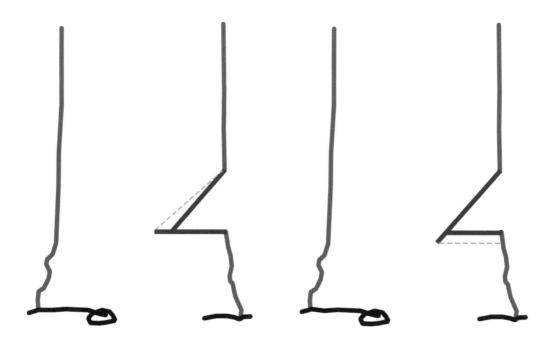

Figure 32: Correction of undercutting. The dotted lines indicate corrective action.

Where the top cut is placed too high, so that continuing the cuts to a point where they intersect would result in the scarf being too deep (see Figure 33), the faller may split the scarf block out leaving a step in the back of the scarf. Provided that both cuts extend into the tree the same distance, this is an acceptable method and is preferred rather than extending the scarf past half of the tree diameter.

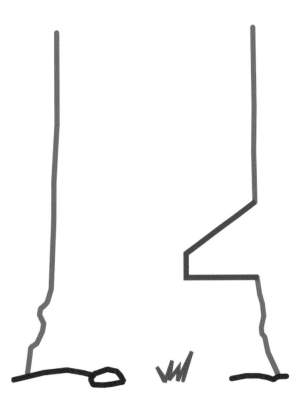

Figure 33: If the intersection of the scarf cuts would be too deep, leave a step in the back of the scarf.

Back cut

The back cut releases the tree from the stump:

- As a general rule the height of the back cut should be a minimum of 50 mm or approximately $\frac{1}{10}$ of the diameter of the tree above the scarf line.
- The height of the back cut provides a step which prevents the tree slipping backwards over the stump (especially when uphill felling).
- Trees may be harder to fell with high back cuts as this creates more hingewood to be broken (particularly dangerous when wedging a slightly backwards-leaning tree).
- The line of the back cut should be level.
- Leave sufficient thickness of hingewood (approximately $\frac{1}{10}$ of the diameter of the tree) to guide tree through intended fell.

Back cut safety

- Back cut should always be completed on the safest side of the tree.
- Avoid walking across the back of a tree to access the escape route.
- Preferred escape route should be on the safest side of the tree.

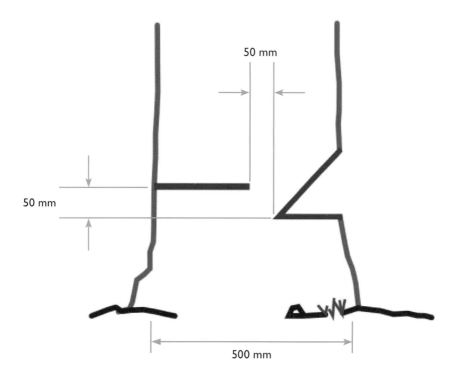

Figure 34: Example of back cut specifications.

Back cutting techniques

The back cut should be above the scarf line and be level. It should finish parallel to the scarf line and leave sufficient hingewood (that section immediately behind the scarf that remains uncut or intact) to provide control. As a general guide the height of the back cut and thickness of the hingewood is about $\frac{1}{10}$ of the tree's diameter – for example, a tree of 300 mm diameter would require approximately 30 mm hingewood (both height and thickness). The height and back cut should be relevant to the size of the tree ($\frac{1}{10}$).

Two common types of back cut are described below.

Method A

This method is used when the tree's diameter is less than the bar length.

The tree is cut through to the hingewood by commencing at the back of the tree. Stand on the tension or opposite side to the lean or crown weight, which is also your escape route side. Cut must be level and at the correct height and must stop when the correct amount of hingewood is achieved. If the tree does not fall, then use a wedge.

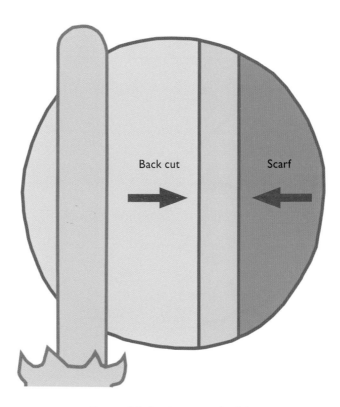

Figure 35: Back cut method A.

Method B

This method is used when the tree's diameter is greater than the bar length.

The tree is cut by boring into the side of the tree under the weight and lean, and at the opposite side to the nominated escape route, moving the bar forward towards the scarf line to set up the required amount of hingewood.

Only about half the tree's diameter should be cut. This can easily be determined by placing the bar along the scarf line and identifying halfway. The saw is then moved in a circular motion, rotating the saw on the bar's nose, finishing up on the opposite side of the tree leaving the correct amount of hingewood uncut.

If there is any doubt about the tree falling, wedges should be placed after completing approximately $\frac{2}{3}$ of the back cut. Once the estimated hingewood has been achieved, check for any reason why the tree has not fallen. If the tree does not fall, do not keep cutting. Refer to the section on lifting trees.

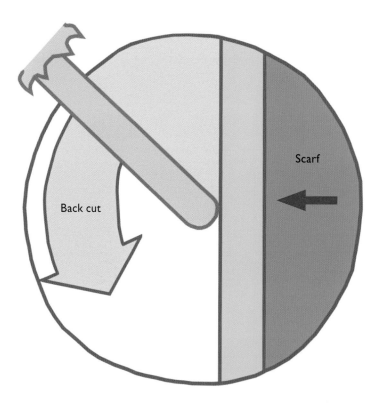

Figure 36: Back cut method B.

Hingewood

A critical component of effective tree felling is the hingewood that is created by the scarf and back cut in combination. The primary control of the tree being felled is achieved by correct hingewood.

The hingewood, as the name implies, is the physical material that holds the tree system to the stump and creates a hinging system as the tree falls.

The principles of the hinge are similar to the many hinge systems we see in a day-to-day environment. Take the example of a car bonnet; this has a hinge placement and action that will allow the bonnet to be opened or closed with confidence, and the hinge will allow precision of position every time. We are looking to achieve this precision with the hingewood in a tree-felling process. In the car bonnet example, as with a tree, if the hinge system is poorly placed, damaged or lacking in strength then the result is at least random. Lack of control in the process of tree felling creates high risk and potentially fatal consequences.

As with many specifications for tree felling, hingewood specifications are affected by the type and species of tree. Generally this relates to the fibre characteristics of the timber in the tree being felled. Seek expert advice on characteristics when attempting to fell different tree species for the first time. A guide to adequate hingewood is to retain $\frac{1}{10}$ of the diameter of the tree.

The characteristics of a good hinge are:

- sufficient fibre is held evenly across the width of the tree, to prevent the tree from twisting or breaking sideways when felling
- adequate strength so that in the event of the tree leaning backwards and closing the back cut it will not break and release the tree in the wrong direction.

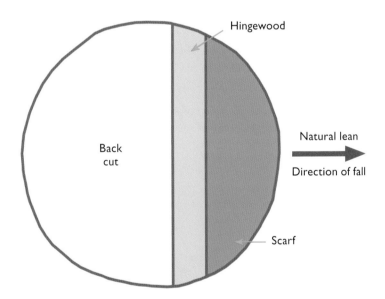

Figure 37: Using even hingewood when felling with the lean.

Hingewood for the side lean

While the general principle applied to hingewood is to achieve even thickness across the tree, in cases where there is side lean present, especially in larger trees, it may be necessary to use tapered hingewood (see Figure 41). With tapered hingewood the operator is trying to hold more wood on the off lean side to counter the pull created by the lean. This practice is limited by the fibre characteristics of the tree, and in severe side lean in many cases it may not be physically possible to hold enough fibre. Care needs to be taken when trying to taper hingewood so that the back cut is not cut too far on the narrow side creating a loss of control or perhaps jamming of the saw.

Splitting trees

One of the high-risk reactions of a tree being felled is splitting of the stem, which in severe cases will create what is sometimes known as a 'barber's chair' and is a potentially deadly situation.

Tree species

Trees may range from plantation pine to numerous native and imported species.

All species have a risk of splitting if not assessed prior to felling and if suitable techniques are not applied. For example, mountain ash *(Eucalyptus regnans)* is well known for its dangerous ability to split very freely if not felled using the correct technique, while slow-grown pine is unlikely to split.

The tree faller must assess the species to be felled for splitting potential before commencing. If the faller is not familiar with the species' characteristics, they should seek professional advice.

Causes of splitting

There are many causes for trees to split. As mentioned above, some species are more prone than others. Splitting may be caused by structural damage in the tree stem but there are many occasions when the pre-felling assessment and cutting technique used are not appropriate for the tree.

Trees with forward lean may split due to the release of tension in the process of back cutting. On other occasions it is poor scarfing procedures that can cause extremely dangerous outcomes.

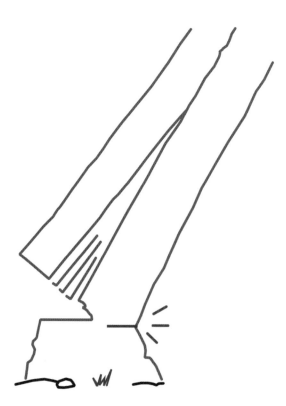

Figure 38: Example of splitting tree.

Felling procedures to minimise splitting

- Visually assess tree species and any indication of stress.
- Visually assess weight distribution and predominant lean.
- Visually assess strength and direction of wind.
- Choose relevant scarfing and back cutting technique for the situation.
- Cut scarf with sufficient angle of opening – greater than 30° with no overcut (see strap technique).
- Back cut accurately and to specification.
- Leave correct amount of hinge – if it is too thick it will split.

Lifting trees

A common problem in tree felling is where the tree does not fall of its own accord due to weather conditions and/or the lean, shape and weight distribution of the tree. Unless full assessment of the situation is undertaken and tools utilised to counter these factors it is inevitable that the operator will have the saw jammed in the back cut at some point. This then poses a significant risk to the faller, chainsaw and equipment.

There is a wide range of tools available to assist the safe felling of the tree, including wedges (plastic and metal) and felling bars, along with tools used only by tree specialists such as tree jacks and winching apparatus.

The wedge is the most commonly used lifting tool for tree felling. When assessing the tree to be felled, the operator should plan to use lifting tools if there is any doubt that the tree will not fall naturally. It is often too late or a high-risk activity to place a wedge once the tree has moved.

The operator must also recognise the limits of lifting tools and appreciate that if the backward lean is too great or there are side lean factors the lift may not be effective.

Wedges

Wedges are most often used when:

- felling slightly backward-leaning trees
- felling side lean trees
- cross-cutting
- heading
- preventing saw jamming
- removing jammed saw.

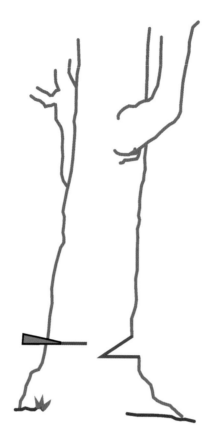

Figure 39: Wedge driven into back cut.

Types of wedges

Wedges come in a variety of materials:

Plastic

- light
- easy on chain (chain will cut through plastic)
- more prone to damage from axe/hammer blows
- some types will 'pop' out if struck hard.

Aluminium

- lighter than steel
- generally won't ruin chain if chain contacts wedge
- less durable than steel.

Steel

- durable but heavy
- will ruin chain if it contacts wedge
- generally used to wedge large trees
- wedge should be driven by sledge-hammer.

Apart from the material chosen, wedge length and lift should be chosen to suit application (tree species, tree size, etc.).

Using a wedge

The type, shape, number and size of wedges to be used will depend on:

- how much lift is required
- how solid the wood is
- how big the tree is.

Placement of the wedge will depend on the desired direction of fall in comparison to the lean of the tree.

Short plastic wedges are good for small trees where a low level of lift is required and the tree has not already sat back and closed the back cut.

Longer and wider plastic wedges are easier to drive in increased lift situations or where the wood is a bit soft or spongy.

Aluminium wedges are more versatile in most lift situations but, because their thickness increases along the length of the wedge more quickly than plastic wedges, are harder to drive in heavy lift situations.

On side-leaning trees the wedge should be placed in the back cut at an angle of approximately 45° to the scarf line to help lift the tree away from its side lean and direct the tree forward to the desired direction of fall.

On backward-leaning trees the wedge should be placed in the back cut at 180° to the desired direction of fall. If using more than one wedge in a backward-leaning tree, the wedges should be spread far enough to allow one wedge to be driven without striking the second wedge with any part of the driving tool.

To facilitate safe and efficient wedging, wedges should be placed in the back cut as soon as there is sufficient room for the wedge not to come in contact with the saw and driven in as the back cut is continued. Heavy or thick bark may have to be removed to allow the wedge to work on solid timber.

Felling bars

Felling bars are tools that often come as a wedge/cant hook combination. These are primarily used for lifting of small trees up to 30 cm in diameter.

Felling bars are used on small trees:

- as an alternative lifting method
- with backward lean
- with some side lean.

As the tree falls

When the tree begins to fall (back cut opens wider than saw cut):

- Withdraw the saw.
- Retreat at least 6 metres along nominated and prepared escape route. Avoid walking across the back of the stump by ensuring the final back cut is completed from the same side as the escape route.
- Continuously watch the tree fall, checking for flying/falling objects.

Post-felling checks

Do not enter felling site until:

- all movement has ceased and the fallen tree is stable
- crowns of neighbouring trees are checked for broken or hanging branches.

SAFETY NOTE
Some trees may be too dangerous to fell by the use of a hand-held chainsaw. The decision as to whether a tree can be safely felled ultimately rests with the tree faller.

The following pages show tree-felling techniques for:

- tree with side lean
- tree with forward lean
- double leaders
- small trees with backward lean
- small tree with the use of wedge or felling bar.

Tree with side lean

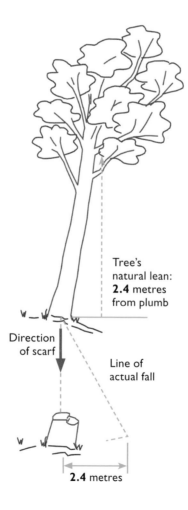

Tree's natural lean: **2.4** metres from plumb

Direction of scarf

Line of actual fall

2.4 metres

Figure 40: Tree with side lean.

When felling a side-leaning tree the faller must take into consideration the distance the head of the tree is from the vertical, as this distance will be translated to a similar distance from the intended direction of fall when the tree hits the ground.

To ensure the head of the tree lands in the desired position the faller will have to scarf the tree further around than the desired direction.

When a tree has to be felled at an angle to its natural lean, there are two main techniques that can be used:

1. tapered hingewood
2. tapered hingewood plus wedge.

Tapered hingewood

- Place scarf in desired direction of fall, taking into consideration the side lean.
- Start cutting back cut on side of natural lean.
- Continue back cut towards opposite side of natural lean, leaving thicker holding wood on that side. The thicker hingewood will help to maintain felling direction through the arc of the fell by giving greater strength to the hinge.
- Start on compression side and finish on tension side.
- Fell to finish back cut on side of tree with the thickest hingewood (safe side of tree).

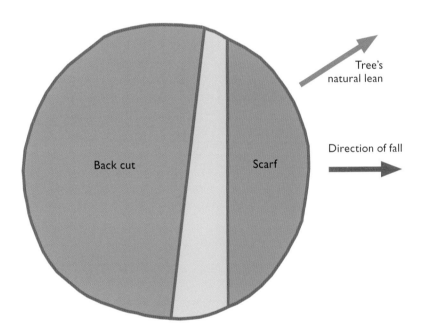

Figure 41: Using tapered hingewood to counter side lean.

Tapered hingewood plus wedge

The technique of tapered hingewood can be assisted by using a wedge in the back cut on the side of the lean.

• Wedge is inserted in the back cut as soon as possible and driven in gradually as the back cut proceeds.

• Wedge should be placed halfway between the hingewood and the back of the tree.

• The thicker hinge will help to maintain felling direction through the arc of fell by giving greater strength to the hinge.

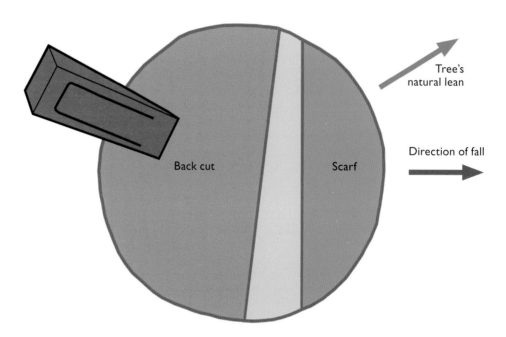

Figure 42: Tapered hingewood plus wedge.

Tree with forward lean

Trees with a forward lean, or free-splitting characteristics, must be felled with caution. These trees may split (barber's chair) if normal felling techniques are used.

> ## This is potentially a very dangerous situation.

The method relies on the faller cutting as much as possible through the middle of the tree before the final release back cut is made. See Figure 43.

A strap is used to hold the tree until the faller is ready to release it and move down the escape route.

Straps are used for three reasons:

- They increase faller safety in difficult situations.
- Some trees will fall faster than the chainsaw can cut (tree may tend to fall over before the hingewood is set up and the tree will split).
- They improve utilisation of the resource by minimising pulled wood.

Back release cut – strap technique

- Standard scarf (**1**) – large as possible without jamming the saw (between $\frac{1}{3}$ and $\frac{1}{4}$ of tree diameter).
- Bore in behind where hingewood is to remain.
- Cut forward to hingewood (**2**), then backward to leave holding strap (anchor) – the strap thickness needs to be 'just enough' to hold the tree.
- Place release back cut (**3**) below boring cut to prevent saw getting caught on 'step' as log falls (12 mm to 25 mm below bore cut/side cut).

Be aware that the tree will fall quickly once the strap is cut.

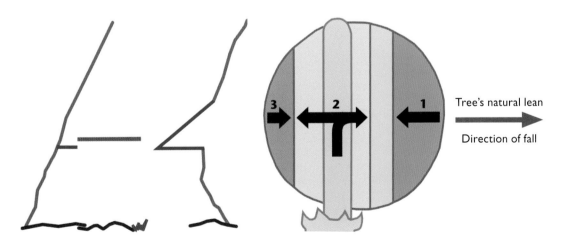

Figure 43: Back release cut.

Double leaders

Double leaders are often difficult to fell. It mainly depends upon where the fork begins. Frequently, even though the fork is fairly high, a weakness runs down some distance below the bottom of the fork.

Where the fork is reasonably close to the ground, and if the tree is felled as a single tree, there is a risk that the leaders may split apart and fall in different directions.

If the fork is high enough, fell as a single tree, generally at right angles to the fork and with the scarf as low as is practicable to avoid the weakness area below the fork.

If it is desired to fell both leaders separately below the fork line:

- rip the joint between the trees to ensure they are separated and so they fall away from each other
- fell the most hazardous leader first
- bore in to begin back cut, consider applying the strap technique as for leaning trees as each stem is likely to be leaning as one grows away from the other seeking light.

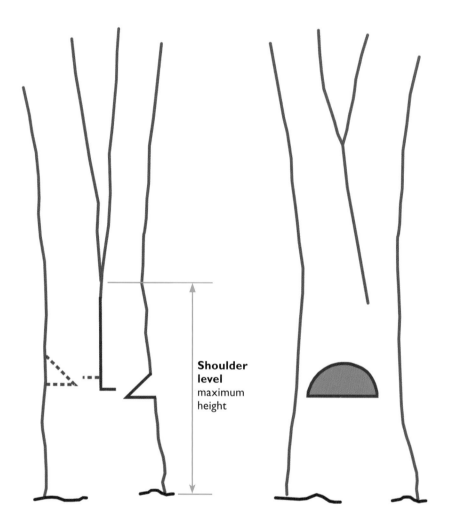

Shoulder level maximum height

Figure 44: Double leaders.

Trees with backward lean

Upon inspection it may appear that the tree is leaning backward or it is not clear that the stem has a forward lean. This may mean that the tree will not fall without mechanical assistance. The operator should not underestimate the dynamics and pressures created by this situation. It could mean that if normal cuts are made, the tree may move in the wrong direction to that proposed and will, in fact, increase its backward lean.

There are two significant risks associated with this:

- If the hinge is not strong enough it may break away and allow uncontrolled fall of the tree (i.e. in an unintended direction).
- If the faller attempts to physically push the tree over there is significant risk of muscular/skeletal injuries (particularly back injuries).

If the pre-fell assessment indicates there is some doubt about the lean or there is obvious backward lean, it is important to still apply correct scarfing principles and use wedges to lift the tree over effectively.

Techniques for small trees with the use of wedge or felling bar

Small trees, generally because of their dimensions, do not allow adequate space to fit a wedge or felling bar in the back cut with the cutter bar. Therefore alternative methods for back cutting are required. See diagrams below.

Technique 1

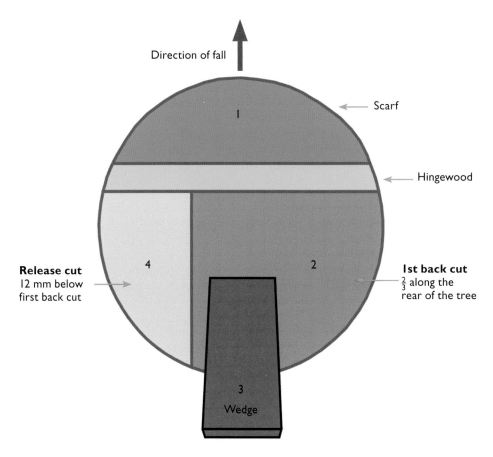

Figure 45: Trees with backward lean.

- Scarf (**1**) the tree with shallower than normal scarf to allow maximum depth of back cut for leverage and room for wedge.
- Partial back cut should always be cut on the lean side first. The initial back cut (**2**) will only cover $\frac{1}{2}$ to $\frac{2}{3}$ of the diameter of the tree with the normal principles of even hingewood and stepping of the back cut.
- Remove any heavy bark to assist with wedging.
- Insert the wedge (**3**) or felling bar firmly, directly in line with the direction of fall (as illustrated).
- Place release cut (**4**), starting at rear of the tree and cutting towards proposed hingewood.
- Overlap cut slightly to ensure all back cut fibre is released.
- Angle second back cut (release cut) slightly (see Figure 45) so that outer part of cut is at the height or slightly higher than the first back cut and the cut in the centre of the tree is slightly below the line of the first back cut. The width of the hingewood should be standard for tree type.

- Avoid matching the two back cuts. If they meet there is a potential for damage to wedge/felling bar and/or saw chain.
- Drive wedge or lift felling bar for tree to fall.

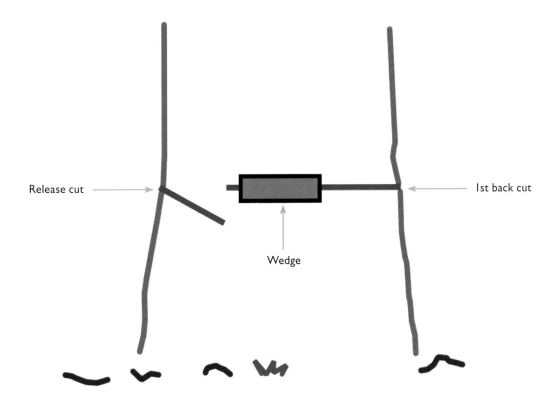

Figure 46: Trees with backward lean.

Note: Trees with significant backward lean will not be able to be lifted with wedges and this should not be attempted. Reconsider direction of fall or other solutions should be sought.

Technique 2

- Scarf tree with shallower than normal scarf, ensuring that the scarf covers $\frac{2}{3}$ of the width of the tree when viewed from in front of the tree.
- While standing at the front of the tree, bore from the scarf through to the back of the tree at the same level as the bottom cut ensuring hingewood is maintained either side of the cut.
- Walk around to the back of the tree and drive the wedge into the bore cut.
- Place two back cuts in the tree either side of and just slightly above the bore cut.
- Continue to drive the wedge into the bore cut as you continue the back cuts until the tree can be wedged over.

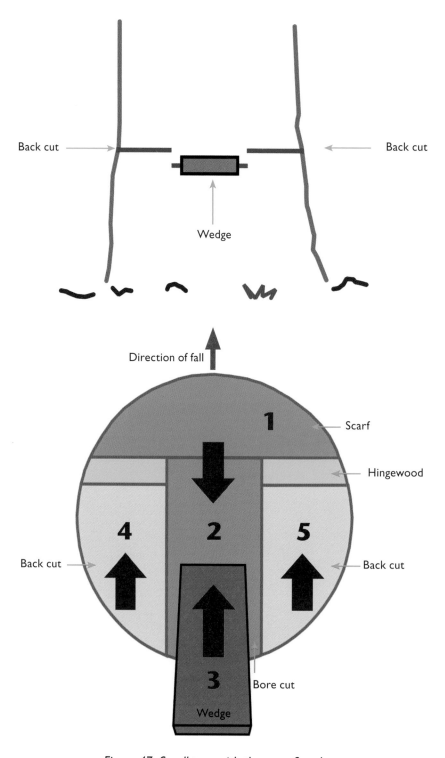

Figure 47: Small tree with the use of wedge.

Complex felling techniques

This section covers the techniques used in the felling of complex trees in plantations and native forest – softwood and hardwood – as well as felling individual hazardous trees.

This advanced felling section of the manual is designed to cover felling techniques for types of trees not previously covered in this resource and which are often referred to as problem trees, although this is not necessarily the case.

It is not possible in this manual to provide advice for every tree-felling scenario and it is always the responsibility of any person who is undertaking the felling of a tree to decide whether they are capable of felling the tree safely. If not, they must be prepared to walk away or seek further advice from other competent tree fallers.

This section will provide advice for felling techniques for the following tree-felling scenarios:

Large	Trees with a diameter up to 2½ times the chainsaw guide bar length
Dead	Trees which appear reasonably sound but have no foliage and may contain dry limbs that may break off during the felling process
Hollow or rotten	Trees that are in an advanced state of decay at the height which the felling cuts are to be made but may contain merchantable wood above the decay
Stag	Trees that may be dead or alive but in an advanced state of decay
Burnt out	Trees that have sustained significant fire damage at the height at which the felling cuts are to be made, and often do not have wood around the entire circumference of the butt, or are standing on legs or spurs
Trees with multiple leaders	Trees that have three or more leaders and may require leaders to be felled individually or the tree to be felled as a whole
Windblown	Trees that have blown over and broken off but are still attached to a vertical section of the stump
Heavy forward leaner	Trees that have a heavy forward lean and have a diameter greater than the bar length

8. Felling hazardous trees

General guidelines

- Faller must be well experienced and have 'Advanced Tree Felling' accreditation.
- Faller must be willing to perform the felling operation and must consider that the felling of the hazardous tree is within their competence level.
- No faller is to be directed to fell a dangerous tree which, in the faller's opinion, is beyond their level of competence.
- If there is not enough daylight to clearly see the site and the drop zone of the tree being felled, then the faller must wait until there is enough daylight.
- Manual tree fallers should have another person checking on their welfare from outside the tree's 'drop zone' while felling hazardous trees.

Large trees

Large trees are trees with a diameter up to 2½ times the chainsaw guide bar length. These can be cut using a centre scarf.

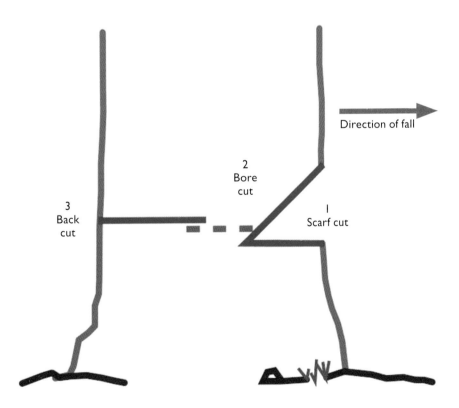

Figure 48: Centre scarf.

Centre scarf

- Make a boring cut into the centre of the scarf at the height of the intended back cut. Make this sufficiently wide so that the guide bar can reach into the cut from either side when putting in the back cut.
- Ensure that there is adequate hingewood on both sides of centre scarf.
- Commence back cut away from safe side of tree and bore in and set up hingewood. Cut around the tree completing hingewood on safest side.
  As hingewood doesn't run all the way across the stump, it therefore must be thicker than normally prescribed. Consider thicker hingewood to compensate for what has been lost by centre scarfing. The use of a V scarf is also an option.

Note: Cuts (**2**) and (**3**) should be on the same level.

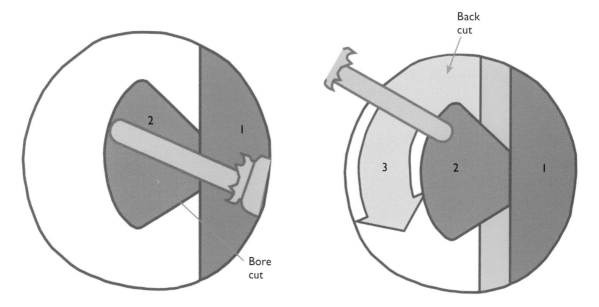

Figure 49: Centre scarf with back cut.

Defective trees

Defective trees are those trees that are unsound, rotten or burnt and may include hollow butts and multi-legged trees. Use the same principles for scarfing, hinge and back cuts as for solid trees. If the tree is rotten or unsound and the wood may not extend across the entire scarf line, the hingewood should be proportionally thicker or higher.

Hollow trees

Hollow trees are those trees that have a pithy/rotten centre or heart or have been burnt out. In any event, any wood that is present is unsound.

- Make a larger than normal scarf but ensure that the scarf meets $\frac{2}{3}$ across the front of the tree. It is critical that there is ample hingewood on both sides to prevent the tree from collapsing.
- If there is enough solid wood, place the back cut in the normal manner but leave more hingewood on either side.
- If in doubt, bore in from either side leaving a solid hinge and cut backwards leaving a strap of hingewood at the rear. (Same technique as for a forward-leaning tree.)

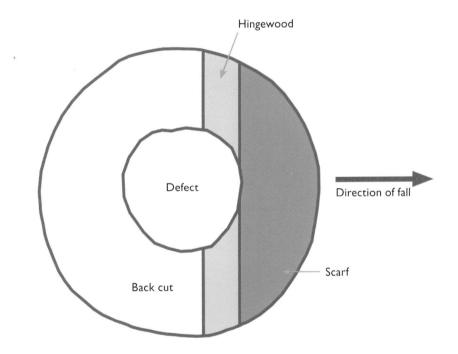

Figure 50: Hollow tree.

Stags

Stags are trees that may be dead or alive but in an advanced state of decay. They generally have no weight in the crown and may have many overhead hazards such as dry limbs.

- The depth of the scarf is increased (up to 50% of the tree's diameter in reasonably sound trees). Consider using the centre scarf technique to reduce the amount of hingewood that the tree needs to break.

- Place back cut lower than normal but still above the scarf to reduce the height of the hingewood.

- Dead trees should not be wedged over unless absolutely necessary. Any wedging should be done with caution.

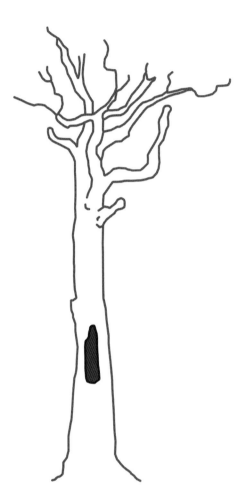

Figure 51: Stag.

**Note: Accurate assessment of defective trees
is critical to faller's safety.**

**Only competent fallers, who have been trained in the felling
of defective trees, should carry out this type of work.**

Burnt-out trees

Burnt-out trees are trees that have had internal wood destroyed by fire and may be standing on legs or spurs.

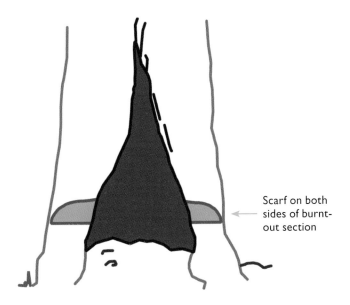

Scarf on both sides of burnt-out section

Figure 52: Burnt-out tree scarf.

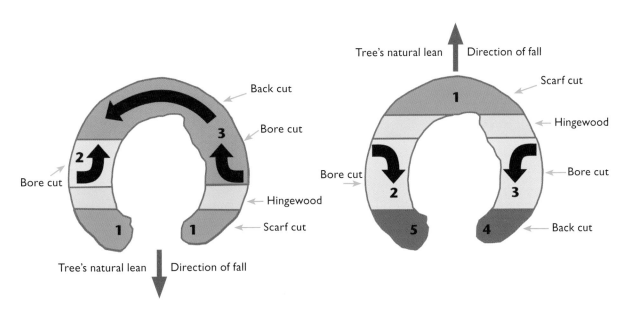

Figure 53: Alternative cutting technique for burnt-out trees.

- Hingewood should be approximately $\frac{1}{10}$ of a tree's diameter when felling in direction of lean – thicker if tree is defective (cross-section through position of scarf).
- Increase the thickness of the hingewood if wood is unsound.

Large multiple leader trees

Multiple leader trees should only be felled by an experienced faller with the relevant knowledge (height, slope, diameter, etc.) to cut the tree or individual leaders.

Multiple leaders may be:

- a co-dominant trunk
- several trunks from the one rootstock
- a split of a dominant leader plus an attached branch
- limbs located close to the proposed cutting position.

These trees will require the faller to accurately assess whether the tree can be felled as a whole tree. Alternatively, the tree may be treated as individual trees and felled using appropriate cutting techniques for each stem.

It may be that even the most competent tree faller has to 'walk away' from this type of tree.

Windblown trees

Felling technique for windblown trees still attached to stump

- Felling this type of tree with a chainsaw may be dangerous.
- Safest method is to push the tree over using a tractor or other suitable machinery.

Windblown or storm damaged trees pose a significant risk to the faller. These trees are usually under extreme pressure or tension. Before attempting to cut trees in these conditions it is important to assess whether they can be tackled manually or if the use of machinery is warranted. Common types of wind-damaged trees include those broken or snapped off above the ground, blown out of the ground with a root ball exposed or bent/stressed.

Tree snapped off above the ground

- The approach to these will depend upon the height at which the breakage has occurred. Anything above shoulder height should be treated as a hanging tree and dealt with using caution.
- Do not attempt to fell the butt/stem. If heavy equipment is available to lift or dislodge the top of the tree it should be used as a priority.
- Undertake a risk assessment that includes:
  - height of break
  - stability of tree stem
  - impact on others
  - tension on stem and branches
  - other related hazards.

If deemed stable then the best approach is to commence cutting at the head of the tree and working towards the butt of the tree.

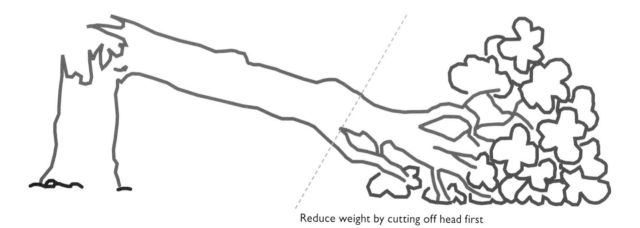

Reduce weight by cutting off head first

Figure 54: Windblown tree snapped off above ground.

Root ball is lifted

Commonly in windblown areas, trees may be blown out of the ground exposing the root plate.

- Again, care must be taken in this situation when processing the tree as there is a significant risk of the root ball rolling or falling back into its original position when the tree is cut. This is potentially dangerous to the faller or bystanders.
- Ensure that the root ball is downhill of the faller and other personnel and will not roll onto the faller.
- If machinery is available it is preferable to stabilise or support the root ball to minimise movement.

Bent or stressed trees

Trees that are windblown and bent over create another dangerous situation for the faller.

A careful risk assessment should be made by asking the following questions:

- How much stress is in the tree?
- Can it be dealt with mechanically?
- Are there issues with the surroundings?

Be very aware that invariably these stems will have significant tension on the top side which may cause the stem to react with unexpected force. Avoid using conventional felling cuts; in many circumstances the back release cut (strap technique) can be used to release these trees. Do not attempt to fell these trees unless the tensions have been thoroughly assessed and evaluated.

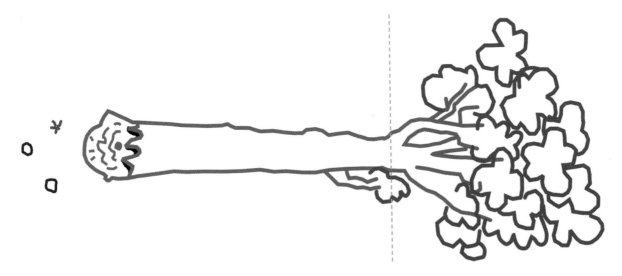

Figure 55: Windblown tree snapped off at ground level.

Heavy forward-leaning trees with diameter greater than bar length

The method used to fell heavy forward-leaning trees with a diameter greater than the bar length is similar to that used to fell smaller trees with a diameter less than the cutter bar length.

Strap technique

1. Cut a standard scarf that may be shallower than normal so the tree will not lean forward, and jam the saw in the cut. The type and age of the tree will be a factor that needs careful assessment in this situation.

2. On the least safe side of the tree bore in behind the proposed hingewood at the height of a normal back cut and cut forward to set up the required hingewood, then backwards leaving a small section of uncut wood (strap) at the back of the tree. The least safe side may be the downhill side, the side with the least amount of solid wood, the side away from your preferred escape route, etc.

3. On the second side of the tree repeat step 2.

4. Cut strap approximately 50 mm below bore cuts to allow tree to fall.

5. Proceed along the preferred escape route (on the same side you were standing to cut the strap). **Be aware**, the tree will fall quickly and may shake limbs loose when you cut the strap, so look up for falling debris.

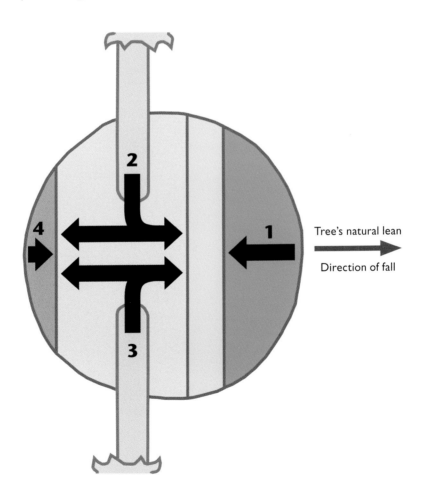

Figure 56: Heavy forward-leaning trees.

9. Other techniques

Machine-assisted manual felling

Machine-assisted manual felling is an advanced method of felling trees against their lean. It is performed with the assistance of preferably an excavator or possibly a skidder, dozer or wheel loader to push the tree over, once the felling cuts have been made, and is quite a common method in production forestry.

This method requires an extremely competent and experienced faller and an equally competent and experienced machine operator, both equipped with electronic communication.

Given that the machinery operation component of this method is quite involved and probably more important than the felling component, this method is considered outside the scope of this manual.

Similarly, line pull felling is also considered outside the scope of this manual.

Tree jacking

Tree jacking is an advanced method of felling trees against their lean. It involves the use of purpose-built hydraulic jacks and a second person to operate the tree jacks.

Once again, because of the significant involvement of the jack operator in this method and the need for advanced mathematical calculations required to determine whether a tree can be safely felled with the assistance of the tree jacks, this method is considered outside the scope of this manual.

Glossary

Back cut	The cut that releases the tree from the stump
Barber's chair	Splitting of the stem
Chaps	Cut-resistant leg protection
Coupe	An area of forest of variable size, shape and orientation, on which harvesting takes place
Crown	The leafy section of the tree
Directional felling	Felling a tree in a specific and identified direction
Drop zone	The area in which a fallen tree will land. For safety the distance is measured at two times the height of the tree from the stump
Escape route	A 6 metre line of retreat 45° diagonally backwards away from the direction of the fall of the tree
Felling bar	A tool that often comes as a wedge/cant hook combination
Gunning sight	Mark on chainsaw cover to assist in achieving desired direction of the fall of the tree
Hanger	Loose or dead branch hanging in a tree that may be easily dislodged
Hingewood	The section of tree between the scarf and back cut that remains uncut or intact and is about $\frac{1}{10}$ of the tree's diameter and acts as a hinge
Hung-up trees	Trees that have lodged together after felling
Intergrowth	Trees that grow within adjoining trees
Lifting tools	Wedges (plastic and metal) and felling bars
Merchantable	Saleable – of quality and type ordinarily acceptable among vendors and buyers
Overcutting	Where one of the scarf cuts extends further into the tree than the other and they do not meet
Plumb bob	A weight that is suspended from a string and used as a vertical reference line
PPE	Personal protective equipment
Release cut	The back cut
Risk assessment	Making an assessment of the risks from the hazards identified
Scarf	Wedge of timber removed from the front of the tree and facing the desired direction of the fall of the tree
Scarf line	The line where the two scarf cuts meet
Sight line	See Gunning sight
Stags	Trees dead or alive but in advanced state of decay
Strap	A strap is uncut wood in the back cut and is used to hold a tree until the faller is ready to release it
Undercutting	Where one of the scarf cuts extends further into the tree than the other and they do not meet
Widow makers	Loose or dead branch hanging in a tree that may be easily dislodged

Relevant state authorities and technical standards

Regulatory and licensing requirements for chainsaw operators and tree fallers vary from state to state. Check with your local state authority listed below.

Tasmania

Regulatory Authority:

Workplace Standards Tasmania

30 Gordon's Hill Rd

Rosny Park,

Hobart Tas 7013

Phone: 03 6233 7676

www.wst.tas.gov.au

Licensing Authority:

Tasmanian Forest Industries Training Board

Shop 4, Cornwall Square Transit Centre

Cimitiere Street

PO Box 2146

Launceston Tas 7250

Phone: 03 6331 6077

Northern Territory

WorkSafe NT

First Floor, Darwin Plaza Building

41 Smith Street

The Mall

Darwin NT 0801

Phone: 1800 019 115

www.worksafe.nt.gov.au

Victoria

WorkSafe Victoria

222 Exhibition Street

Melbourne Vic 3000

Phone: 1800 136 089

www.workcover.vic.gov.au

Western Australia

WorkSafe WA

5th Floor, Westcentre

1260 Hay Street

West Perth WA 6005

Phone: 1300 307 877

www.commerce.wa.gov.au/WorkSafe

Queensland

Workplace Health and Safety Queensland

75 William Street

Brisbane Qld 4000

Phone: 1300 369 915

www.deir.qld.gov.au/workplace/

New South Wales

WorkCover NSW

92–100 Donnison Street

Gosford NSW 2250

Phone: 02 4321 5000

www.workcover.nsw.gov.au

South Australia

SafeWork SA

Level 3,

1 Richmond Road, Keswick

GPO Box 465

Adelaide SA 5001

Phone: 08 8303 0400 or 1300365 255 (within SA)

www.safework.sa.gov.au

Technical standards

AS 2727 – 1997	*Chainsaws – Guide to Safe Working Practices*
AS 2726.1 – 2004	*Chainsaws – Safety Requirements, Part 1: Chainsaws for General Use*
AS/NZS 4453.3 – 1997	*Protective Clothing for Users of Hand-held Chainsaws, Part 3: Protective Legwear*
AS/NZS 1801 – 1997	*Occupational Protective Helmets*
AS/NZS 2210.1 – 1994	*Occupational Protective Footwear, Part 1: Guide to Selection, Care and Use*
AS/NZS 1337 – 1992	*Eye Protectors for Industrial Applications*
AS/NZS 1336 – 1997	*Recommended Practices for Occupational Eye Protection*
AS/NZS 1319 – 1994	*Safety Signs for the Occupational Environment*
AS/NZS 1270 – 2002	*Acoustics – Hearing Protectors*
AS/NZS 4602 – 1999	*High Visibility Safety Garments*